Greville Macdonald

On the respiratory functions of the nose

Greville Macdonald

On the respiratory functions of the nose

ISBN/EAN: 9783337196998

Printed in Europe, USA, Canada, Australia, Japan

Cover: Foto ©berggeist007 / pixelio.de

More available books at **www.hansebooks.com**

ON THE

RESPIRATORY FUNCTIONS
OF THE NOSE

AND

THEIR RELATION TO CERTAIN
PATHOLOGICAL CONDITIONS

BY

GREVILLE MACDONALD, M.D. (LOND.)

PHYSICIAN TO THE THROAT HOSPITAL, GOLDEN SQUARE, W.

LONDON
ALEXANDER P. WATT
2 PATERNOSTER SQUARE
1889

PREFACE.

On the eve of publication Dr. E. Bloch, of Freiburg (Baden), kindly sends me a pamphlet, by himself, and published in translation in 'The Archives of Otology' (New York), vol. xvii., No. 4, 1888. It is entitled 'Physiological Investigations of Nasal Respiration.' I regret very much not having seen this before, if only to give Dr. Bloch credit for his very painstaking work. At the same time I cannot but be gratified in discovering that his methods and results closely correspond with my own. Dr. Bloch's experiments are more elaborate, embracing the thermometric and hygrometric conditions of the nasal air, as well as the filtering functions of the nose. The latter point I have not discussed. On the other hand, Dr. Bloch does not refer to gaseous exchanges occurring in the nasal chambers; nor does he touch upon the structure and functions of the inferior turbinated body, or inquire into the relation of the physics to the pathology of the nose.

The results of my experiments on the thermometry and hygrometry were communicated to the British Medical Association at its annual meeting in Glasgow, July, 1888, and published in the journal, Dec. 1, 1888.

G. McD.

QUEEN ANNE STREET,
July 17, 1889.

CONTENTS.

THE

RESPIRATORY FUNCTIONS OF THE NOSE.

IT is commonly held, and has been taught
for many years, that the nose is the natural
respirator—in other words, that so long as
breathing through the nose is possible there
can be no need for any apparatus covering
the mouth to warm, filter, and moisten the
inspired current of air. This doctrine is now
considered scientific by most medical men,
and consequently the artificial respirator is
almost discarded, although not so completely
as one might desire. Nevertheless, the true
importance of nasal respiration in the economy
has hitherto not received the attention it

deserves, and experimental physiology has almost neglected the comparatively simple nature of investigations on the functions of the nose. Aschenbrandt of Wurzburg has made some investigations as to the temperature and hygrometric condition of nasal air which closely agree in their results with mine ; but his method is at once less easy and less trustworthy, while some recent repetitions of his experiments have led to unsatisfactory results.* Naturally, the pharynx, larynx, and trachea are the first organs to suffer from obstructed nasal respiration, seeing that upon these organs devolve nasal duties for which they are not specially adapted ; and laryngologists are generally teaching that the nose is of paramount importance in the effectual treatment of many cases coming under their special notice. Yet, such observations have been

* 'Ueber die Bedeutung der Nase im Respiration.' Wurzburg, 1886.

based on a theory of nasal breathing rather than on physical fact.

In the following pages I propose attempting, in the first place, to give the nose a firmer position in physiological science, relating in detail experiments that have occupied a large portion of my time for the last twelve months; in the second place, to inquire into the structure and functions of the inferior turbinated body; and, lastly, to indicate the relation that the physiology of the nose bears to certain diseases with which we are clinically acquainted.

Before describing the experiments, it will be not inappropriate to acknowledge my indebtedness to Mr. Sidney Skinner, B.A., of Christ's College, Cambridge, for most courteously assisting and supervising my calculations in various deductions; to Mr. Kauffmann, Superintendent of the Linde Refrigerating Works, for the great kindness he extended to me in permitting the use of an atmosphere artificially reduced in temperature; and to my brother, Mackay MacDonald, of Christ's College, Cambridge, for much assistance in carrying out the investigations.

I. Experimental Physiology of the Nasal Respiratory Functions.

In order that a tracheotomized patient may not be deprived of the benefits of nasal respiration, it occurred to me that by connecting a piece of indiarubber tubing at one end with the tracheal cannula, and at the other with one nostril, the inspired current of air would be compelled to enter at the unattached nostril, traverse the nasal cavities on both sides by passing round the septum, exit at the other nostril, and descend by the tubing to the windpipe. In this way the air would undergo its natural modification in the nose, and the patient consequently be placed less at a disadvantage by his operation. In many particulars, however, this proved to be im-

practicable ; but it suggested a simple
method of obtaining for analysis air that has
passed only through the chambers of the
nose without encountering the mucous membrane of the pharynx and mouth, and un-
contaminated with air from the lungs. And
upon this principle all my experiments have
been conducted, although, as a matter of fact,
but few on a tracheotomized subject. I found
that the same course was pursued by the
current of air passing in at one nostril and
out at the other, if the tubing was connected
with the mouth instead of with a tracheal
cannula. The very attempt to inhale through
the tubing rather than through the nose is
sufficient to raise instinctively the soft palate,
and thus close the post-nasal space. That
this is a correct statement will be satisfactorily established when the experiments
are discussed in detail. Consequently, it is
competent for any person capable of breathing naturally through the nose to perform

the experiments, the apparatus for some of them being so simple that it may be purchased for a very few pence.

Before proceeding, however, I may call attention at the outset to an objection that will probably suggest itself as a serious obstacle. It is this : that the air passing in at one nostril and out at the other has traversed double the natural area of mucous membrane, and, therefore, that the experiments are not trustworthy. The answer to this argument is this : that if the air be drawn through the nose by a natural act of respiration, as already suggested, passing in only at one nostril, it will necessarily traverse that with double the velocity attained when breathing is performed through both air-passages. Consequently, it remains in contact with the mucous membrane only half the time, although by passing through both sides it traverses double the area ; and we may assume that the one condition balances the other. As a matter of fact, moreover, the real objection to my method of experimenting is in the reverse direction. One cannot be sure that the current of air has sufficient access to the walls of the post-nasal space. In all probability this region bears a very small part in these experiments, the work of the nose being consequently under-estimated rather than the reverse. And this difficulty I see at present no method of overcoming. Be this as it may, the experiments are sufficiently convincing.

For convenience' sake, I will divide the investigations into three headings: (1) the degree to which the temperature of the inspired air is raised by the nose; (2) the degree of humidity acquired under the same circumstances; and (3) the chemical changes that take place in the air by passing through the nose alone.

1. The Degree to which the Tempera-ture of the Air is raised by the Nose.

THE method in which this investigation was conducted is as follows. An ordinary piece of **T** glass tubing is taken and one arm bent round till the extremity lies beneath the vertical portion (see Fig. 1). At the angle of junction a small bulb is then blown ; into the vertical portion is inserted a thermometer graduated from 60° to 100° F. (15° C. to 37° C.), carrying an index, and fixed into the tube with a collar of rubber tubing (A). The bulb of the thermometer must fall well into that of the outer tube, but not so far as to come into contact with it. To the straight and bent arms are then respectively fixed a solid rubber perforated nipple (B), large enough to fit tightly into one nostril, and a piece of tubing (C) of sufficient length to be held between the lips when the instrument is

in situ. The index then having been shaken
down below 70° F. (20° C.), the whole of the

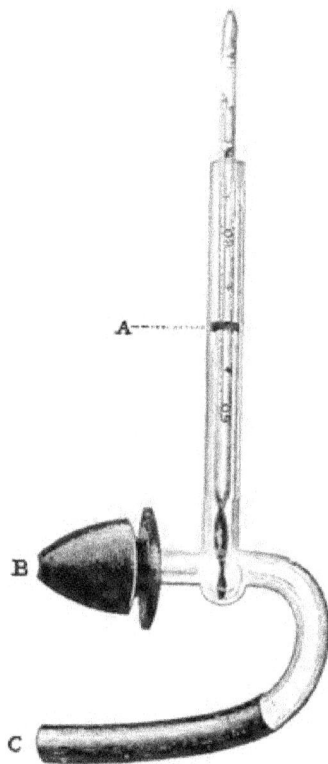

FIG. I.

A. *Rubber Collar.*
B. *Nosepiece.*
C. *Mouthpiece.*

apparatus is covered with Berlin wool, or
enclosed in a flannel jacket, so as to exclude

all sources of heat except that of the air passing through the tube. Next an observation is taken of the atmospheric temperature, and, after a full expiratory effort, the apparatus is fixed with the nipple (B) in one nostril and the rubber tubing (C) in the mouth. Through the latter the air is inspired, and is immediately felt rushing in at the open nostril. That no air passes from the nose into the pharynx is proved by closing the rubber tube with the finger and thumb, when respiration is immediately stopped, it thus being manifest that the inspired air traverses only the area comprised in the two nasal fossæ. Expiration must be performed through the mouth while the mouthpiece is removed, the latter being tightly held, not only to exclude cold air, but as an additional safeguard against expired air passing through the nose and so over the thermometer. I employed at first a valvular arrangement that automatically permitted of expiration without removal of the mouth-piece ; but the arrangement described is both simpler and securer. Four to six deep inspirations generally suffice to raise the thermometer to the highest point ; but with the lower temperatures under which the experiment has been conducted, double the number may be necessary. The length of time necessary for exposure necessarily varies with the sensitiveness of the thermometer.

The following are a few results casually selected from a great number of experiments on different individuals. In the left-hand column the atmospheric temperature is given, and in the right that to which the air is raised after passing through the nose :

At $-7°$ C. the temperature was raised to $28·8°$ C.
At $1·7°$ C.   ,,   ,,   ,, $35°$ C.
At $7°$ C.   ,,   ,,   ,, $34°$ C.
At $12°$ C.   ,,   ,,   ,, $35·6°$ C.
At $45°$ C. the temperature was reduced to $33·6°$ C.

A consideration of this table indicates the remarkable fact that, at any rate between the above thermometric limits, whatever the atmospheric temperature, the inspired current of air on passing through the nose alone is raised or lowered in temperature, approximately to that of the blood. And, in all probability, the results of the experiments would be far more striking were it possible to include the mucous membrane of the post-nasal chamber in the area traversed by the inspired current.

The experiments with the lowest atmospheric temperature quoted were conducted at Linde's Refrigerating Works, Deptford. In this connection it is interesting to note that in the same temperature, *i.e.*, $-7°$ C., air breathed in and out of the lungs only, without the intervention of the nose, was raised to $33·5°$ C.; whereas when breathing was conducted in at the nose and out at the mouth, the thermometer indicated $35°$ C., the duration of the respiratory acts occupying the same number of seconds in each case. This observation corresponds

2

with a statement of Gréhant, to the effect that the temperature of the expired air, when breathing was conducted through the mouth, indicated 33·5° C., while with nose breathing it showed 35·2° C.*

The experiments with an atmospheric temperature of 45° C. were conducted by passing the inspired air through a 30-foot composition gas-pipe coiled round in boiling water, a thermometer being inserted in this circuit immediately before its termination in the nostril. The absorption of aqueous vapour in the nose renders latent so much heat that the thermometer indicates actually a lowering by three degrees of the temperature of the air below that of the blood.

Ceteris paribus, a robust subject raises the temperature two, three, or even more degrees

* 'Recherches Physiques sur la Respiration de l'Homme': *Journal de l'Anatomie et de la Physiologie*, 1864. See also experiments of Sir Morell Mackenzie, 'Diseases of the Throat and Nose,' vol. ii., p. 372.

than an anæmic. In the latter we generally have more or less collapse of the erectile tissue lining the inferior meatus. A noteworthy fact, moreover, is that cocaine, by anæmizing and inducing collapse of this tissue, lessens the acquired temperature by two or three degrees. These points would appear to throw considerable light on the functions of the inferior turbinated bodies, although the small amount of difference in the observed temperatures in erection and collapse of these structures indicates perhaps that they do no more than share the function of heating with the rest of the mucous membrane.

These experiments establish unequivocally the fact that the nose alone is competent to raise the temperature of the inspired air, whatever its degree, almost, if not quite, to the temperature of the blood.

2. THE DEGREE OF HUMIDITY ACQUIRED BY THE AIR IN PASSING THROUGH THE NOSE.

The experiments falling under this head were conducted in a manner similar to those just described in ascertaining the temperature of air passed through the nose, but with the addition of more complicated apparatus. It was necessary to pass a known volume of dry air in at one nostril, and on its exit from the other to collect the aqueous vapour absorbed by it. A reference to the accompanying woodcut (Fig. 2) will make a description more intelligible. An ordinary bell-jar (C), measuring four litres, its upper opening closed with a caoutchouc stopper carrying a piece of glass tube, is connected by flexible tubing with a large calcium-chloride vessel, and this in a similar manner with one nostril. A tube making its exit from the other nostril conducts the air through a series of three other calcium-chloride vessels, the last of

which is once more attached to a tube con-
veying the air into the mouth. The act of

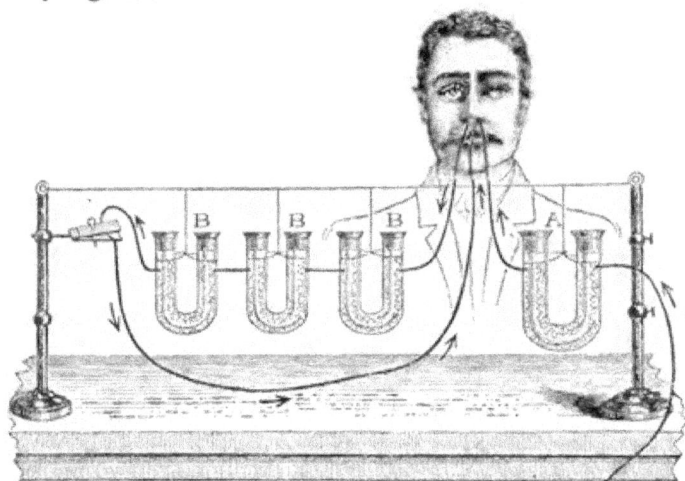

FIG. 2.

A. }
B. } *Calcium-chloride Tubes.*
C. *Bell-jar to measure Air.*

inspiration through the mouthpiece necessarily draws air out at one nostril, in at the

other, and finally exhausts the bell-jar. The
latter is held by an assistant during the experiment over a bath of water sufficiently
deep to allow of its complete submersion,
and is allowed slowly to descend into the
water as the air is exhausted ; thus the
volume of air respired is accurately measured.
Immediately before the experiment is made,
the calcium-tubes (B, B, B) are weighed
either together or severally. They are then
replaced in the scheme, the tubes are so
firmly fixed in the nose as to admit of no
leakage, the mouthpiece is secured, and the
assistant holds the bell-jar in contact with
the surface of the water. A slow, deep
inspiration through the mouthpiece gradually
exhausts the bell-jar as it descends in the
water. As soon as the thorax is full,
the tubes entering the nose are firmly
grasped, and that in the mouth removed and
closed, before expiration is permitted through
the· latter channel. The process is then re-

peated until the receiver is exhausted of air and filled with water. Twenty litres of air should be thus passed through the nose before the calcium-tubes are again weighed. The difference in the two observations, allowance being made for the increase of volume due to the rise in temperature, will thus give the amount of water absorbed by twenty litres of dry air on passing through the chambers of the nose.

That the inhaled air actually pursues the course described is manifested at any moment by compressing digitally any one of the rubber tubes. If, for instance, that by which the air exits from the nose be closed, respiration is immediately arrested, showing that no air enters the thorax by the other two tubes. And in the same way every point of the scheme may be investigated. But, to make conviction doubly sure, valves may be placed in the tubes in such a manner as to prevent the passage of air in any direction but that desired—a precaution obviously quite unnecessary.

The following are a few instances, among many, of the results of passing twenty litres of dried air through the nose. In the first column the subject experimented on is mentioned. A was in perfect health, B slightly anæmic ; in both the nasal passages were fairly normal in their patency—a point the importance of which will be established later. In the second and third columns are mentioned respectively the atmospheric temperature and that to which the air was raised after passing through the nose. In the fourth column are given the differences in the weighings before and after the experiment, these representing the number of grammes of water absorbed.

SUBJECT, ETC.	TEMPERA-TURE OF AT-MOSPHERE	TEMPERA-TURE AFTER PASSING NOSE	GRAMMES OF H_2O ABSORBED BY 20 LITRES OF DRY AIR
A.	17·7° C.	34·3° C.	1·170
B.	17·7° C.	33·8° C.	1 033
A.	20° C.	35° C.	1·024
B.	20° C.	34·2° C.	0·775
A.	21° C.	35·3° C.	1·963
A. with the erectile tissue collapsed under cocaine.	20° C.	34·4° C.	0·681

In order to indicate the true importance of these figures, it is necessary to determine the amount of water absorbed by twenty litres of air when completely saturated. This is estimated in the following formula :

$$\cdot08936 \times 9 \times V \times \frac{273}{273+t} \times \frac{P}{B}.$$

t = temperature at which volume is measured.

B = barometric pressure.

P = vapour pressure of water at tempera-ture of nose (41·8 at 30°C.).

V = volume of air.

$$\text{Or } \cdot 08936 \times 9 \times 20 \times \frac{273}{273 + 16} \times \frac{41 \cdot 8}{760} = \cdot 835$$

grammes.

This calculation, however, is based on the assumption that the barometric pressure within the nose is 760 mm. Thus ·835 grammes of H_2O are contained in twenty litres of water, when saturated, at the temperature of the nose—an amount actually less than the experimental observations quoted above. Practically, however, the very fact that the inspiratory act consists in the formation of a vacuum within the thorax, and consequently in all the passages leading to it, clearly indicates that the barometric pressure within the nose during inspiration must be less than that of the external atmosphere. The lower the air tension, the greater is the amount of aqueous vapour absorbed, *ceteris paribus.* And consequently it is only to be supposed that air at the intra nasal pressure during inspiration would contain at saturation a

greater quantity of water than at the external barometric pressure. And this the
above observations prove to be the case.

Other experiments, to be described later,
where the current of air was passed through
fine potash-tubes, and the difficulty of inhaling thereby greatly increased—or, in other
words, where there was necessarily an increased degree of vacuum—indicated an
enormous addition to the amount of water
absorbed.

In one of these experiments, for instance,
twenty litres absorbed 1·371 grammes of water,
almost double that of saturation at 760 mm. ;
while a control-experiment made immediately
afterwards, the potash-bulbs being excluded
from the scheme, gave only ·519 grammes.
The latter very small amount was probably
accounted for on the supposition that the
mucous membrane was exhausted by its
previous artificially accelerated secretion.

From a consideration of the physics of

respiration and the laws of gaseous mechanics,
we must consequently assume that the amount
of moisture taken up by the inspired air varies
to a certain extent (1) with the rapidity of the
inspiratory act, (2) with the degree of turges-
cence of the erectile tissue in the nose, and
(3) with the calibre of the nasal passages
—a point which varies considerably with
the physiognomy of the individual, etc.—
or with morbid conditions of partial nasal
obstruction.   But to these points I shall
have to refer again, when speaking of the
relation of the physics of the nose to the
production of certain diseases in that organ.
Yet, in spite of such sources of variation, we
may assume that the air in passing through
the nose alone is completely saturated with
moisture, and that consequently physiologists
are in error when they tell us that the lungs
exhale moisture.   This cannot be when the
air reaches them already saturated.   Nay,
more : at the moment the expiratory act

begins, and the barometric pressure in the
lungs exceeds, instead of being less than, that
of the atmosphere, the air may be supersaturated; and, remarkable as it may appear,
we may have, to a certain extent, a deposition
of dew in the air-passages.

Again, when we remember the enormous
amount of heat rendered latent by the
vaporizing of water, we are struck with
astonishment at the large quantity im-
parted by the nose, not only in raising the
temperature of the inspired air, but even
more in charging it with moisture.

3. The Chemical Changes that take place in the Air in passing through the Nose.

We now pass on to the third series of experiments, viz., those for determining the chemical changes that take place in the air as it passes through the nose.

It is doubtless generally assumed that interchange between oxygen and carbonic acid occurs to a certain extent over all mucous surfaces exposed to the atmosphere. But, so far as I am aware, this has not hitherto been submitted to direct investigation. The following experiments were undertaken not only with the object of establishing the fact, but of determining the actual quantity of gaseous interchange as well as the variations in this quantity with varying external atmospheric temperature.

To determine the presence of carbonic acid in the air passed through the nose, we adopt a method similar to that described in the previous experiments. The inhaled current of air is drawn first through a Liebig's or Geissler's potash-tube to remove all traces of carbonic acid as it exists in the atmosphere ; it then passes into a solution of baryta to indicate its freedom from the gas, and finally into the nose. As the air emerges from the other nostril, it passes again into baryta, out of which it pursues its course to the mouth and lungs. The arrangement is shown in Fig. 3, the vessels A and B containing the baryta - water, and C the potash solution. Before placing the tubes in the nose and mouth, a few inspirations must be taken through the nose and exhaled from the mouth in order to empty the nasal chambers of any residue of carbonic acid as exhaled from the lungs. After the last such exhalation the tubes are properly ar-

ranged as in the diagram, and a single
inspiration taken through the mouthpiece.
This is sufficient to cloud the solution in
the bottle A, while that in B remains clear.

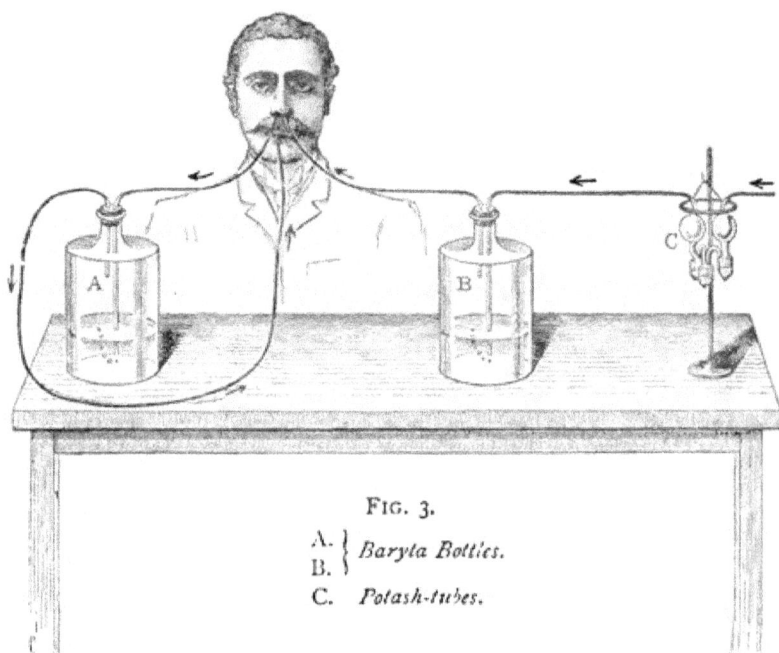

FIG. 3.

A. } *Baryta Bottles.*
B. }
C.   *Potash-tubes.*

As the air is drawn through by a single
inspiratory act, it is obviously impossible for
any air from the lungs to vitiate the observation.

This   simple   experiment   incontestably

proves the fact that exchange of gases takes place to a certain extent in the nose just as in the lungs.

The next experiment is quantitative in the same direction. For this the apparatus is again more complicated, and the process more tedious. It is similar to that used for estimating the hygrometric condition, with the omission of the first chloride of calcium tube, as it is not necessary to dry the air before its entrance into the nose. In its place we have a Geissler's potash-tube, A, Fig. 4, and between this and the nostril an empty tube C, to collect any solution of potash that might otherwise enter the nose. As the air makes its exit from the nose, the acquired water is removed by the tubes B, B, B, containing pure chloride of calcium. Two of these large tubes are sufficient to remove all the water. Many of the experiments, however, were made with first two and then four tubes

3

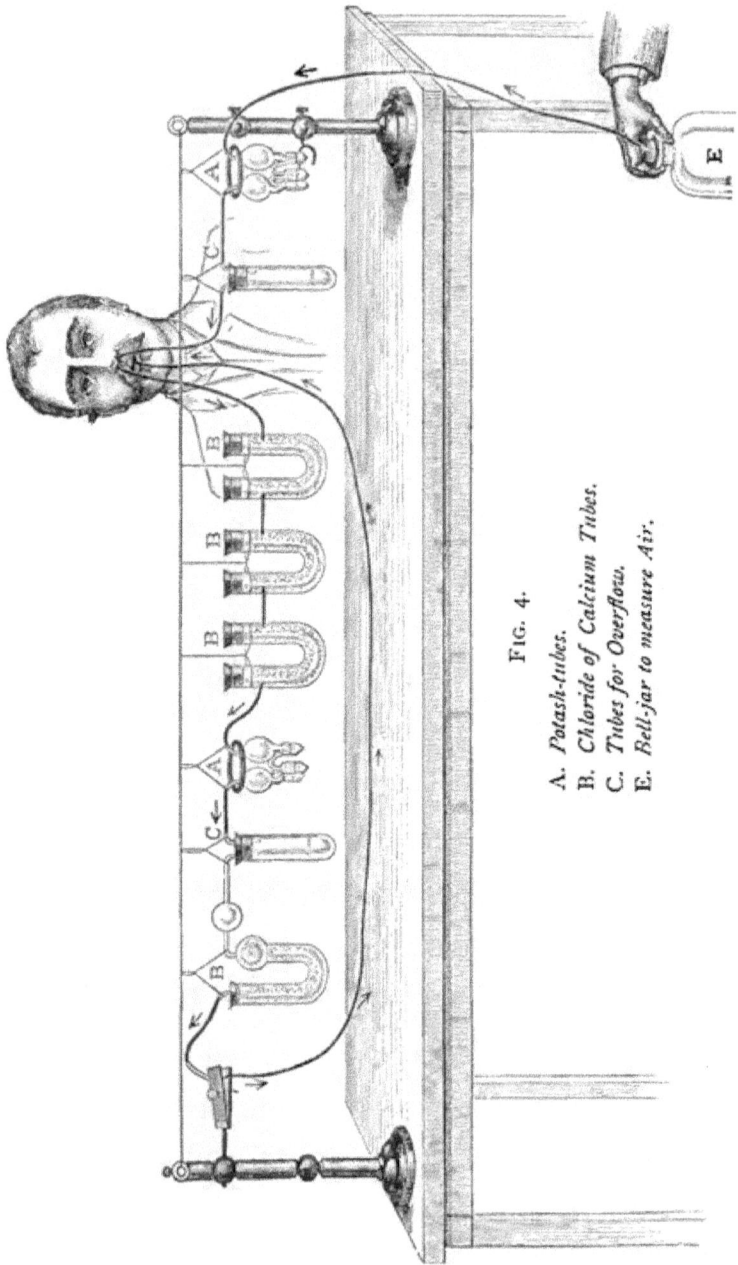

FIG. 4.

A. *Potash-tubes.*
B. *Chloride of Calcium Tubes.*
C. *Tubes for Overflow.*
E. *Bell-jar to measure Air.*

without altering the result. Three was the usual number employed. The current then passes into potash again, through another tube for collecting overflow, and once more through a single chloride of calcium tube, the purpose of which is to retain any moisture abstracted by the dry air from the potash-solution. Finally the air passes by a long piece of tubing into the mouth and lungs.

All the precautions of the former experiments must be religiously observed: (1) After each inhalation the tubes must be removed from the nose and mouth before expiration takes place; (2) the latter must be performed only through the mouth; (3) the nose must be well emptied of accidental carbonic acid from the lungs by a series of inspirations through the nose and expirations through the mouth before the experiment is begun. With care in these directions, it will be impossible for carbonic acid from the lungs to complicate the experiment. Inhaling the air through this complex arrangement of tubes, many of them necessarily of fine bore, demands considerably more muscular effort than is employed in ordinary breathing. The assistant controlling the bell-jar (E) can facilitate inspiration considerably by forcibly depressing the jar, and so driving the air onwards. A firm, steady hand is essential for the success of this procedure.

Before beginning the actual experiment, the final set of potash-tubes, etc. (A, C, and B), must be carefully weighed : the increase of weight after the experiment will show the amount of carbonic acid contained in a known volume of pure air after passing over the nasal mucous membrane. It was found sufficient to pass twenty litres through the nose, this involving the filling of the bell-jar five times.

Similar experiments were made at various atmospheric temperatures to determine whether the amount of carbonic acid given off by the nose is in any way commensurate with the different amount of heat contributed to the air in the different cases.

The extremes of temperature experimented with in this manner were 1·7° C. and 38° C. Many of the observations at these temperatures were made consecutively, that for the lower being conducted in a cold atmosphere, while the air was artificially heated for the experiment immediately preceding or following. The method adopted for the latter purpose was as follows : A spirit-lamp was placed beneath a glass tube conducting the air to the nose, but separated by some inches of rubber tubing from a T tube, in the upright arm of which was fastened a thermometer, the proximal end of this tube being fixed in the nostril. The reading of the thermometer thus gave the temperature of the air immediately before entering the nose. In making the calculations, allowance

was of course made in every case for the
increase of volume in the air due to its rise

NO. OF EXPERI-MENT	SUBJECT	ATMOSPHERIC TEMPERA-TURE	GRAMMES OF CO_2 GIVEN BY NOSE TO 20 LITRES OF AIR	CONDITIONS
1.	A.	15° C.	·033	
2.	A.	14·5° C.	·034	
3.	A.	15° C.	·058	Face flushed and circulation quickened with alcohol.
4.	A.	ditto.	·023	Two hours later, face and circulation normal.
5.	A.	13° C.	·033 }	Immediately consecutive experiments.
6.	A.	1·7° C.	·041 }	
7.	A.	1·5° C.	·041	
8.	A.	1·7° C.	·042 }	Immediately consecutive.
9.	A.	21° C.	·020 }	
10.	A.	1·7° C.	·033	
11.	A.	10° C.	·022	
12.	A.	30° C.	·017	
13.	B.	1·7° C.	·039 }	Consecutive.
14.	B.	27° C.	·032 }	
15.	B.	4·5° C.	·041 }	Consecutive.
16.	B.	37·5° C.	·030 }	

in temperature, either by the spirit-lamp, or
on passing through the nose.

Experiments 3 and 4 were performed with

four chloride of calcium tubes through which
the air passed after its exit from the nose,
and before entering the potash-tubes.

The deductions from these experiments are
interesting as adding additional weight to the
former qualitative experiment with baryta-
water, all proving that, to a certain extent, the
nose performs precisely the same function as
the lungs themselves. Yet beyond this remarkable fact it will prove instructive to inquire further, (1) exactly what proportion
of carbonic acid transudes through the nasal
mucous membrane in proportion to that
thrown off by the lungs; and (2) the relation
which the amount of carbonic acid formed in
the nose bears to the quantity of heat
yielded to the inspired air under varying
atmospheric temperatures.

1. Assuming that the average proportion
of carbonic acid in expired air is 4·38 per
cent., the figures usually quoted in the physiological text-books, and that one litre of car-

bonic acid weighs 1·977 grammes, a hundred
litres of expired air would contain 8·66
grammes of carbonic acid, and twenty litres
1·73 grammes. Now, the average quantity
of carbonic acid contained in twenty litres
of air passed through the nose is about ·033
grammes; or, in other words, nearly $\frac{1}{50}$th of
the exhaled carbonic acid is given off by the
nasal mucous membrane, even when expiration is not performed through the nose.
Consequently in ordinary respiration, with
the mouth closed, possibly even more than
this quantity is contributed by the nose. But
in determining this ratio we must bear in
mind a possible source of fallacy already
briefly referred to. This is found in the fact
that, with the thoracic effort to inspire through
the complicated system of tubes above described, the barometric pressure within the
nose is necessarily unnaturally low; or, in
other words, an unusual degree of vacuum is
created in the thorax in order that the ex-

ternal pressure may be sufficiently raised to
force the air. through the scheme of tubes.
This low degree of air-tension, again, within
the nose must produce to a certain extent not
only an over-filling of the capillaries and increased blood-supply, but also a heightened
tension of the gases in the blood, and so
augmented transudation into the nose.
Against this, however, it may be remarked
that, while the bell-jar is forced down into the
water at the beginning of each descent, the
pressure is sufficiently raised within the
scheme of tubes to force the air through
the nose without any inspiratory effort whatever. This fact obviously counterbalances to
a certain extent the liability to error just mentioned. But the quantitative import of such
sources of fallacy I have not attempted to
realize.

2. In determining the relation which the
amount of carbonic acid formed in the nose
bears to the quantity of heat yielded to the

inspired air, we are dealing with more precise equivalents. And on calculating the amount of carbon which it is necessary to convert into carbonic acid in order to raise twenty litres of air 20°C., we are struck with the close approximation of the result with that of the experiments.* But in this consideration it must be remembered that no account has been taken of the heat rendered latent by absorption of aqueous vapour in the nose, a factor varying within very wide limits accord-

---

* To estimate grammes of carbon to be burnt to raise 20 litres of air 20°C. :

Sp. heat of air is ·2374, or ·2374 heat-units raise one gramme of air to 1°C.

To raise 20 litres of air 20°C. $= 20 \times 1·293 \times 20 \times ·2374.$

Now, the combustion of 1 gramme of carbon gives 8,000 heat-units, and

$$\frac{20 \times 1·293 \times 20 \times ·2374}{8000} = ·013$$

represents grammes of carbon necessary to raise 20 litres of air 20°C.

Finally, ·013 grammes of carbon represent ·047 grammes of carbonic acid, a close approximation to the quantities of that gas obtained on passing 20 litres of air through the nose.

ing to the hygrometric condition of the atmosphere. This latent heat is generally so great in quantity as to justify the assumption that there is not sufficient combustion of carbon within the nasal mucous membrane, etc., to supply all the heat required, and, therefore, that much of the heat is supplied to the air by conduction and radiation. The following questions nevertheless naturally suggest themselves: Is there any special and direct combustion in the tissues of the nose for the purpose of supplying heat to the air, and beyond that which occurs in like tissues elsewhere? or is the inspired air warmed merely by contact with the mucous membrane, the formation of carbonic acid being, so to speak, accidental?

The marked increase of carbonic acid in the nasal air when the external temperature is low would at first sight appear to indicate some special power of controlling oxidation in the nasal tissues, seeing that no such differ-

ence in the composition of air expired from
the lungs has been observed under similarly
varying conditions. But in such a supposition we may be confusing cause with effect.
For in either case we have necessarily an
increased blood-supply, which would not only
keep warm the constantly cooling mucous
membrane, but must obviously lead to increased supply of oxygen to the tissues and
increased formation of carbonic acid. This
tissue-change would in its turn, moreover, lead
to augmented generation of heat.

Consequently we are not advancing an
hypothesis unsupported by fact, if we assume
that, on the contact of colder air with the nasal
mucous membrane, there is an immediate increase in the capillary blood-supply, with
accelerated oxidation and elimination of carbonic acid.

This concludes the account of the experiments, as well as some of their teachings. The latter I may now briefly summarize in a statement as to the physically ascertained functions of the nose, so far as its respiratory functions are concerned :

1. However cold the atmospheric temperature, the air is raised almost, if not quite, to the temperature of the blood, on passing through the nose alone, and before reaching the pharynx.

2. However dry the external air may be, on passing through the nose it is completely saturated with moisture.

3. Gaseous exchanges take place in the nose between the gases of the blood and those of the air, just as in the lungs, and that to a not inconsiderable extent. Moreover, the quantity of carbonic acid exhaled by the nasal mucous membrane is proportionate to the number of degrees in temperature to which the air is raised. This increase in the

heat-supply is probably due partly to increased conduction, radiation, etc., of heat from the augmented blood-supply to the mucous membrane, and partly to direct increase of oxidation in that and the subjacent structures.

II. On the Structure and Function of the Inferior Turbinated Body.

The peculiar erectile property of the mucous membrane covering the inferior spongy bone has been a matter of interest and speculation for many years. But in the light of the important functions performed by the nose, it gains additional interest; and it may not be inopportune in this place to inquire more minutely into its anatomy, its method of tumefaction, and the purpose which it thereby fulfils.

In removing with the cold snare hypertrophied portions of the erectile tissue, especially the enormous growths that occasionally project backwards from the posterior extremity of the inferior turbinated body, I have succeeded in obtaining microscopical specimens in which the vessels are injected with blood. In this condition a very cursory examination is sufficient to manifest the whole structure and mechanism of these bodies.

We find three well-defined layers, (i.) the epithelial, (ii.) the fibro - vascular, and (iii.) the submucous, in which are contained the racemose glands and venous sinuses, to which latter the erectile property is due. These sinuses form a loose spongy network with little connective tissue separating their walls. When distended, the latter are seen to consist of a thin layer of fibrous tissue, apparently not elastic, and lined with an endothelium continuous with the veins which open into them. When empty, the walls become corrugated, and lie in close contact with one another. The arterioles, frequently tortuous, but becoming straightened when the structure is distended with blood or when œdematous, run directly towards the surface, and there ramify in capillary vessels. These are gradually united into the radicle veins, which, in their turn, pursue a more or less direct course towards the venous sinuses, into which they directly empty their contents.

All the specimens I have examined unquestionably reveal this arrangement, and they appear to me to set at rest the long-disputed mechanism. In the fibro-vascular and submucous layers there appears to be a considerable development of elastic tissue, to which is doubtless due the property the venous network possesses of emptying itself when the blood-supply is cut off by contraction of the arterioles. Thus the mechanism of erection and collapse of the inferior turbinated body is quite simple.

Toynbee was probably the first to refer specially to the erectile property possessed by the nasal mucous membrane ;* while

* 'Diseases of the Ear,' 1868, p. 200.

4

Kohlrausch described, as arranged vertically to the bone, venous loops which he injected from the jugular vein.[*]   But to Bigelow, of Boston, is rightly ascribed the merit of discovering the truly erectile property of the cavernous structures of the nose.[†]   He says that the thin trabeculæ and walls, composed mainly of connective-tissue, closely resemble the corporaca vernosa ; although, in the latter, the smooth muscular element, as also the tunica albuginea, is somewhat more pronounced.

I have failed to discover muscular fibres in the trabeculæ of the turbinated bodies, although Bosworth, of New York, subscribes to the statement ;[‡] nor is there much that can be considered as corresponding with the tunica albuginea.   Moreover, the mechanism of erection appears to be different. For although we find tortuous arteries, as I

---

[*] Müller's 'Archiv,' 1853.
[†] *Boston Med. and Surg. Journ.*, 1875.
[‡] *New York Med. Journ.*, 1886, p. 327.

have remarked, yet they are scarcely identical with the helicine arteries of J. Müller, as Bosworth appears to assume,* seeing that they never open into the venous spaces directly. The last-mentioned observer has latterly denied the existence of a venous erectile tissue, though on what grounds is not very clear.† Sajous, of Philadelphia, describes the capillaries as opening abruptly into the venous sinuses,‡ an observation that I am satisfied is not correct.

The function of these erectile bodies appears to be that of increasing the area of mucous membrane over which the air passes, when either more heat or more moisture is demanded. For instance, after exposure to cold we find the bodies assuming their greatest proportions, while in an atmosphere supersaturated with moisture we frequently

---

* Trans. Int. Congr., 1881, vol. iii., p. 327.
† *New York Med. Journ.*, 1886, p. 492.
‡ 'Diseases of the Nose and Throat,' 1881, p. 16.

find them collapsed. This last statement, however, needs some qualification, as the patients on whom the observations were made had been breathing an atmosphere not only foggy, but charged with the irritants of London smoke. Further experiment is needed upon this point.

The mechanism of erection appears to be vaso-motor. Increased blood-supply over-fills capillaries, veins and sinuses. The latter become over-distended, and produce tumefaction of the whole structure. Presumably, also, the filling of the sinuses induces a stagnation of blood in the veins and an increase of pressure in the arterioles, thus augmenting transudation and secretion—a process similar to that obtaining in the kidneys, where free secretion into the Malpighian body is facilitated by the high arterial pressure due to the double system of capillaries through which the blood passes. That some such regulation in the secretion from the turbinated bodies is necessary, one is readily persuaded when he remembers that the colder the atmosphere, *ceteris paribus*, the greater the amount of moisture necessary for saturating the air, raised to the blood-temperature, with aqueous vapour.

No sooner is the arterial supply cut off than the sinuses are enabled to empty themselves, owing to the elastic property of the tissues in which they lie. And it is thus that cocaine, by inducing contraction of the arterioles, cuts off the supply of blood from the sinuses, and so induces collapse of the spongy bodies.

III. The Relation of the Physiology of the Nose to certain Pathological Conditions.

Certain nasal and pharyngeal diseases are more obscure in their etiology and pathology than ordinary inflammatory affections, because in their structure they are so simple that, on anatomical grounds, they can scarcely be considered the products of morbid activity. I refer especially to certain hypertrophies of normal tissues with which, in treating nasal obstruction, we have to deal even more frequently than with ordinary inflammatory products. Such obstructions include (1) post-nasal adenoid growths, (2) true hypertrophy of the inferior turbinated body, (3) ecchondroses, exostoses, and deflections of the septum. With each of these there exists no structural point to distinguish it from the normal tissue in continuity with which it grows. This, I think, will be at once conceded, except, perhaps, for the hypertrophied inferior turbinated body. The actual increase of all the elements of which this latter consists—venous sinuses, fibrous, elastic, and adenoid tissues, glands, etc.—is beyond all question, so far there being no evidence of inflammatory action. The point in which the hypertrophied differs from the natural body is the œdema of the fibro-vascular layers of the mucous membrane. But from a consideration of the physiological conditions of œdema, I believe we have no reason for considering this to be necessarily of an inflammatory nature. This point I shall have to develop further on.

Considering, then, these affections as the result of physiological hypertrophy, the question naturally arises as to whether any special conditions exist in the nose that would induce a supply of blood greater than is needed for the maintenance of the normal nutrition. Why, we naturally ask, should this tendency to hypertrophy be found especially and so frequently in the nose? and that, moreover, in structures so different as bone, cartilage, lymphoid tissue, and venous sinuses? May we not possibly find some common factor that will account for all these changes? This is the point I am venturing now to discuss.

In the first place, a reference may not inappropriately be made to some of the experiments already detailed, and the facts deduced from them.

In the second series of experiments, viz., those for determining the hygrometric condition of the air after passing through the nose, it may be remembered that where, from any cause, we have difficulty in nasal inspiration, the amount of moisture absorbed is considerably increased. This fact was satisfactorily explained when it was remembered that the barometric pressure within the nose is necessarily diminished under the conditions in question, and that, as a physical consequence, the attenuated air would absorb more moisture. These experiments, of course, were not necessary to prove the obvious physical fact; yet they showed in a very forcible manner that, whenever we have nasal stenosis, there must be diminished air-tension behind the seat of stenosis, so long, at any rate, as respiration is conducted through the nose, and not through the mouth.

Now, all fluids, whether liquid or gaseous, obey the same laws, and tend to fill any vacuum with which they are in communication. Hence, not only does air rush in at the nose, but the blood-vessels—arteries, capillaries, and veins—which line the vacuum, become overfilled.* Increased blood-supply

---

* This fact, if not sufficiently obvious, is easily submitted to ocular demonstration. If, with Siegle's pneu-

means increased nutrition, while the latter
necessitates hypertrophy of the tissues
supplied. We know that in children lymphoid tissue, wherever situated, is peculiarly
liable to increase in quantity under slight
provocation ; so that it is quite intelligible
that such structures as the naso-pharyngeal
adenoid tissue, and, perhaps, also the
buccal tonsils, would be among the first
to suffer.*

Under the same conditions it would be
probable that the whole structure of the
inferior turbinated body should become

matic speculum, the air be partially exhausted from the
external auditory meatus, the walls and membrana
tympani are immediately suffu ed with blood.

* Since going to press I find that Bosworth has
insisted upon ' the influence of diminished atmospheric
pressure due to nasal stenosis upon the mucous membrane of the air-passages beyond the point of obstruction, as leading to dilatation of the blood-vessels and
weakness of vaso-motor control, thus giving rise to
attacks of hay fever in the nasal chambers and spas-
modic asthma in the bronchial tubes' (*New York Med.
Journ.*, 1886, p. 492).

hypertrophied in consequence of the increased intra-vascular pressure. The œdema,
moreover, of the fibro-vascular layer of the
mucous membrane may be accounted for in
the same manner. According to Cornil and
Ranvier, the conditions most favourable to
the production of œdema are venous obstruction and increased capillary pressure—the
very conditions, indeed, which necessarily, in
obedience to physical laws, obtain when the
air-tension in the nose is unnaturally reduced.
Venous obstruction is induced owing to the
pressure within the veins exceeding that of
the partial vacuum which they line, while
the intra-capillary pressure is necessarily
increased for precisely the same reason.

It will justly be observed that if inspiration
through a constricted passage produces a
minus pressure, expiration will exert the
opposite effect, and that thus the equilibrium
in the blood-supply will be maintained.
Moreover, it will doubtless be remembered,

as a dictum of physiology, that the air-tension within the respiratory cavities is greater during expiration than inspiration. This, were we discussing perfectly normal states, would destroy the theory I am advancing. But we are dealing, as a matter of fact, with abnormal stenosis, which, as I will immediately show, reverses these conditions of pressure. For the purpose of demonstrating this experimentally, I have constructed an instrument which I may call the naso-manometer (Fig 5). It consists of an ordinary U-shaped mercurial pressure-tube, one arm being conveniently bent, so that it may be firmly fixed in one nostril. Resting on the summit of the mercury in the other arm is an iron weight, attached to a string passing over a pulley, and kept tense by a small weight fastened to its free extremity. The pulley rotates a pointer passing over a dial graduated in magnified millimètres. On attaching the nose-piece of the instru-

ment firmly in one nostril by means of a
rubber nipple, the subject breathing naturally
and easily through the other, the pointer
indicates very precisely the intra-nasal

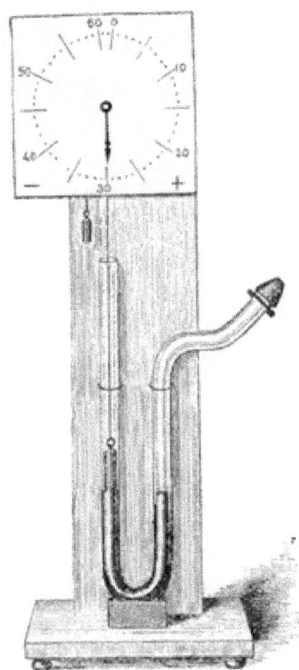

FIG. 5.

pressure of the air as it passes in and out,
and consequently, as accurately, any obstruction to the easy entrance of air. In the
latter case we find that the minus pressure

of inspiration is much greater than the plus pressure of expiration, and that the greater the degree of stenosis, the greater the difference between the observations.*   The same remark applies whether the respiration be easy or forced.   And, as a matter of clinical observation, patients suffering from nasal stenosis frequently volunteer the information that expiration through the nose is so much easier than inspiration.

Donders, who, I find, since writing the above, used a U-shaped manometer fastened in one nostril for determining the pressure in the air-passages, gives us the following as the results of his observations : During quiet breathing inspiration indicated 1 mm. [Hg.], and expiration + 2-3 mm. ; while forced inspiration showed − 57, and expiration + 87. As a matter of experiment, any

---

* This naso-manometer is occasionally of use in determining whether one or other nostril is partially obstructed, the dial indicating the degree of stenosis in the nostril in which the nose-piece is not fixed.

individual, with normal nose-passages, may verify these figures with the instrument above described ; and then, by partly closing the open nostril and so obstructing the access of air, he can substantiate my observations on the pressure conditions in narrowing of the nasal channels.

It is a little difficult to account for this discrepancy in the physiological and patho- logical conditions. That, when there is nasal obstruction, the air-tension should be hardly augmented in easy expiration may be accounted for by the fact that in such a state there is no muscular effort : the air being expelled merely by the elasticity of the chest-walls, the outflow is rather delayed than the pressure of the air increased. Similarly, in forced breathing, the power of the muscles of inspiration greatly exceeds that of the expiratory.

Once more, it may be objected that, in the event of there being any obstruction to easy

breathing through the nose, patients instinctively adopt mouth-breathing. But that this
does not invariably hold in cases of partial
or transient nasal stenosis I have frequently
observed. The mouth may be shut, and
breathing performed through the nose ; yet
we see, during inspiration, that falling in of
the upper half of the anterior triangles, and
the exaggerated descent of the larynx and
trachea towards the thoracic cavity, which
are both so pathognomonic of lessened
atmospheric pressure within the respiratory
chambers. In normal breathing the anterior
triangles do not perceptibly move, and the
descent of the larynx is so slight as to
be scarcely noticeable. Moreover, I have
observed that sometimes during sleep the
instinct of nasal breathing, even when con-
tending with great difficulties, strongly asserts
itself ; though, perhaps, as a rule the patient
is more likely to adopt buccal breathing
during sleep than when awake. Some forms

of snoring, indeed, so far from being the result of the open mouth, are due to the great difficulty with which air is drawn in and out of the nose.

It yet remains to be shown that nasal stenosis of some sort is the usual starting-point of those structural hypertrophies with which we are concerned.

It is a matter of general observation that post-nasal growths are frequently, and in children nearly always, accompanied by some form of nasal stenosis. The more common is vascular tumefaction of the inferior turbinated body, with or without chronic rhinitis. And it frequently, perhaps generally, happens that some treatment is demanded for remedying this, beyond removal of the post-nasal growths, before nasal respiration is completely restored. Seeing, moreover, that these adenoid neoplasms occasionally date from an attack of measles or other exanthem, we may assume that a catarrhal obstruction antecedes the structural. But of more value in throwing light on possible sources of obstruction leading to this hypertrophy is the frequently observed narrowness of the nasal fossæ from congenital osseous malformation. In connection with this we see a highly-arched palate, with or without the V-shaped superior maxilla of the dentists, the short upper lip, and generally narrow physiognomy. And it is probably when cleft-palate is associated with this condition, as frequently happens, that we find also post-nasal growths, as related by Sir Morell Mackenzie, Meyer, and Loewenberg. My observations in this direction have not been sufficiently numerous to be of any value.

In obstructions arising from true hypertrophy of the inferior turbinated body, whether anterior or posterior, we generally have a history of chronic rhinitis, which would, from vascular swelling, induce the primary obstruction.

If we are in any doubt as to the applica-
bility of the theory to anterior enlargements
of the inferior turbinated body, we must remember that we never find hypertrophy in
this situation without the whole length of the
structure being involved to as great, and
generally to a greater, degree. And with
such free intercommunication of the blood-
vessels as we find in this cavernous tissue,
we can scarcely conceive of one portion
suffering without the other. Nevertheless,
we sometimes find posterior hypertrophy with
very little anterior hypertrophy, although invariably with much vascular swelling.

For ecchondroses and exostoses we may
be in some doubt as to whether the theory
will apply, seeing that they are sometimes
situated anteriorly to the transient obstructions of catarrh. As a matter of fact, there
is almost always a history of a blow in those
cases where the deflection and thickening of
the septum is situated in the prominent

portion of the nose. But in the majority of
other cases, the hyperæmia induced by a
low tension of the air in the nasal fossæ will
account for the hypertrophy, and perhaps
also for ossification of the cartilage. The
deflection of the septum, which we generally
see as the concomitant of ecchondrosis, is
explicable upon the same hypothesis. Seeing
that, above and below, the septum lies between fixed limits, any augmentation of
substance in the vertical direction, such as
would occur in a general hypertrophy,
must necessarily bend the structure to one
side or the other. Almost all deflections,
moreover, except those of a traumatic origin
find their most prominent point posteriorly
to a line drawn from the free margin of the
nasal bones to the nasal spine of the superior
maxilla. I am not maintaining that deflection
and ecchondrosis of the septum are always
sequels to an obstructive rhinitis, since
traumatism, as I have already remarked, is

responsible for many cases. But that it often occurs in conjunction with such a rhinitis I am fully persuaded, especially since I have watched such hypertrophies in the process of their growth. And it must have been re-marked by every surgeon who has had a large nose-practice, how frequently obstinate cases of nasal polypus are associated with exaggerated degrees of hypertrophy in both cartilaginous and osseous portions of the septum. Perhaps the œdema of nasal polypus is due partly to the same causes.

It is interesting to note in this connection that while post-nasal growths are almost confined to infancy, the great majority of deflections and thickenings of the septum occur subsequently to puberty.

That the tonsils should enlarge under diminished air-tension is quite conceivable, especially where they exist in conjunction with post-nasal growths and catarrhal nasal stenosis. But of course inflammatory action has generally an important share in such hypertrophies.

Certainly many cases occur, both of post-nasal adenoids and ecchondrosis of the septum, where no mucous or muco-purulent discharge points to present inflammatory action. But in most cases we have more or less hyper-secretion, due either to the hyperæmia or direct irritation of the abnormal products.

It is needless in this place to do more than refer to the pharyngeal and palatal appearances in post-nasal growths. The small growths in the pharynx, and the congested, semi-œdematous condition of the soft palate, are explicable upon the same theory that accounts for the post-nasal adenoids; while the laryngeal and tracheal catarrh in the subjects of these and other nasal obstructions can as readily be explained. Perhaps the reason we see fewer nutritive changes resulting from the abnormal degree of vacuum

external to the bony framework of the skull
is that, lower in the respiratory track, the
walls of the vacuum itself yield to the external atmospheric pressure. This we can
see externally in the neck and thorax
during both pathological and artificial nasal
stenosis.

A few words will suffice for summarizing
the theory I have advanced and endeavoured
to maintain. As a necessary consequence of
physical laws, and from the teaching of direct
experiment, it follows that where we have
obstruction to the free access of air to the
respiratory passages there we have diminished
air-tension. Where the walls of this partial
vacuum are yielding, their collapse probably
tends to minimize the ill-effects that would
otherwise accrue. But where, as in the nose,
the walls are rigid, all the blood-vessels lining
them become over-filled, and a state of hyper-
nutrition is induced, which in its turn tends
to the increase in bulk of one or more of the

tissues so affected. Thus are produced those hypertrophies and œdemas with which the rhinologist is so familiar, but at so much loss satisfactorily to account for.

THE END.

BILLING AND SONS, PRINTERS, GUILDFORD.

-

www.ingramcontent.com/pod-product-compliance
Lightning Source LLC
Chambersburg PA
CBHW020236090426
42735CB00010B/1712